YOUR KNOWLEDGE HAS VALUE

- We will publish your bachelor's and master's thesis, essays and papers

- Your own eBook and book - sold worldwide in all relevant shops

- Earn money with each sale

Upload your text at www.GRIN.com and publish for free

Bibliographic information published by the German National Library:

The German National Library lists this publication in the National Bibliography; detailed bibliographic data are available on the Internet at http://dnb.dnb.de .

Imprint:

Copyright © 2017 GRIN Verlag
Print and binding: Books on Demand GmbH, Norderstedt Germany
ISBN: 9783668656864

This book at GRIN:

https://www.grin.com/document/414394

Hans Georg Schrey

Thermohydraulic Comparison of Fin Tubes

GRIN Verlag

Title   **Thermohydraulic Comparison Procedure of Fin Tubes**

Author   Hans Georg Schrey

Equations: Microsoft formula editor

Figures: EXCEL graphics

On behalf of all authors, the corresponding author states that there is no conflict of interest.

<u>Abstract</u>

Fin tubes are core elements of air cooled heat exchangers in industrial cooling applications. Thermohydraulic performance of the cooling system defines the overall size of the equipment. Fin tube characteristics vary over a large range resulting from geometry, material or manufacturing process. As a general rule - the better the performance the smaller and more competitive the design will be. Consequently, manufacturers strive for thermohydraulic optimization of their product line. A key factor is the ability to properly compare different cooling devices.

Comparison methods have been a topic in the academic field over a long period of time. Methods proposed so far have been based on classical dimensionless parameters, especially Reynolds number. Apart from transport properties Reynolds numbers include a geometric parameter pertinent to the individual heat exchanger system. If geometry is varied Reynolds changes as well. This is one of the reasons why Reynolds based comparison methods – although theoretically sound – lack practical applicability.

The following report attempts to make up for the deficiencies of theoretical Reynolds comparison by using an approach based on redefined parameter groups – specifically meant for cross-flow air coolers and/or air-cooled condensers. It will allow to directly evaluate different fin tube systems based on their performance characteristics alone.

Content

1 Introduction

Atmospheric cooling systems are generally arranged in cross-flow with different number of rows and passes. The cooling medium is ambient air flowing at the external (fin) side by either forced or natural draft. Basically, inlet/outlet planes of air and process sides are oriented perpendicularly whatever the internal row-pass arrangement may be. All types of dry air coolers have this feature in common – irrespective of process medium, tube geometry, single or two-phase heat transfer.

Manufacturers all over the world have developed a multitude of fin tube variations with the focus on cooling effectiveness, low cost manufacturing and material availability. Core tube geometry is one of the noticeable features of different fin tube systems – see fig. 1 with examples of (a) channel type, (b) circular or (c) elliptical core tubes. Other variations are in the form of the finning with single fins or fin sheets covering the bundle. Plain fins are used as well as turbulence promoted surfaces to augment heat transfer. There is also an abundance of fin or tube pitches. Apart from classical multi-row arrangements the bulk of cooling bundles for air-cooled condensers consists nowadays of single-row fin tube bundles with an elongated channel type core tube and only one tube row (fig. 1a). These examples show that the variety of geometry is extreme. Not surprisingly, general correlations covering all these variants are not available. Therefore, it is impossible to define a common Reynolds with only one hydraulic diameter as a comprehensive flow number.

The size of the air side inlet/outlet plane defines the total plot area in each form of bundle arrangement. Apart from that auxiliary power consumption and total surface requirement to fulfil the heat duty define total system cost. The similarity of atmospheric cooling systems with respect to inlet/outlet plane may therefore be used by a general approach.

To ease the optimization the classical form of fin tube characteristics will be modified considering the face area concept. Basically, the alternative form of fin tube correlations has been used by Kroeger [1] and others.

2 Classical Form of Fin Tube Characteristics

Fin tube characteristics are commonly summarized by correlations based on dimensionless numbers of heat transfer and pressure drop coefficient. The typical form is

Reynolds number: $\qquad Re = \rho \cdot u_c \cdot d_h / \eta$

Heat transfer: $\qquad Nu = h_A \cdot d_h / \lambda = C_0 \cdot Re^m \cdot Pr^{1/3}$

Pressure drop: $\qquad \Delta P = \varsigma \cdot \tfrac{1}{2} \cdot \rho \cdot u_c^2 \cdot [l_h / d_h]$

Resistance factor: $\qquad \varsigma = D_0 \cdot Re^{-q}$

As for pressure drop, it is convenient to include the ratio of hydraulic flow length to diameter into the correlation factor. Theoretically, the heat transfer coefficient - based on total fin side surface - must be corrected by a fin efficiency factor. However, in standard atmospheric cooling the change of fin efficiency is small over the typical range of air temperatures. Following general practice this effect may be implemented into the heat transfer coefficient.

The main critical issue of general fin tube correlations is the definition of hydraulic diameter. As mentioned already the complexity of fin tube geometries cannot be covered by one geometric parameter alone. Researchers have circumvented the problem of varying air cross-flow velocity within the structure by using the maximum air velocity within the narrowest flow gap for definition of Reynolds number – so that details such as inter-fin gap, fin height or fin structure are taken into account to at least some extent. On the other hand, the narrowest inter fin gap is normally not used for the definition of hydraulic diameter. The core tube diameter is used in most correlations. So, practically there is a large variation of Reynolds numbers caused by different hydraulic diameter definitions at similar air velocity which makes Reynolds unfit as general operation variable. Therefore, in the proposed alternative classical dimensionless parameters will be modified.

3 New Form of Fin Tube Characteristics

The proposed form of correlations uses a black box principle (fig. 2). For definition of parameters and variables see the attachment. Face area velocity which is independent from the type of fin tube system is used as defining air speed. The area concentration ratio combines face velocity directly to maximum air speed in the narrowest gap. By switching to face velocity the tube bundle internal geometry is made part of the correlation constant. Also, the hydraulic diameter may be defined with reference to face area and in this way is identical for all fin tube variations. Consequently, the hydraulic diameter can be made part of the correlation constant as well.

Thus, the new pseudo Reynolds flow number is defined as

$$Ry \equiv \frac{\rho \cdot u}{\eta} = \frac{Re}{d_h} \cdot \left(\frac{A}{A_c} \right)^{-1} \tag{1}$$

For convenience, the relation to the classical number is given. Heat transfer is also referred to face area. This gives a new heat transfer (pseudo Nusselt) number as

$$Ny \equiv \frac{h_A}{\lambda \cdot Pr^{1/3}} \cdot \left(\frac{F}{A} \right) = \frac{Nu \cdot d_h^{-1}}{Pr^{1/3}} \cdot \left(\frac{F}{A} \right) \quad . \tag{2}$$

The resistance factor correlation takes the following form:

$$\varsigma = D_0 \cdot Ry^{-q} \cdot d_h^{-q} \cdot \left(\frac{A}{A_c} \right)^{-q} = D_0 \cdot Re^{-q}$$

For comparison purposes however the absolute pressure drop is of interest. Leaving out the length to diameter ratio the new pressure drop (pseudo Euler) number is defined as

$$Ey \equiv \frac{\rho \cdot \Delta P}{\eta^2} = \frac{1}{2} \cdot \varsigma \cdot Ry^2 \cdot \left(\frac{A}{A_c} \right)^2 \tag{3}$$

In the new form the fin tube characteristics may be correlated as

$$Ny = C \cdot Ry^{m}$$

$$Ey = D \cdot Ry^{n}$$

Constants of classical and new form are connected in the following way:

$$C = C_0 \cdot d_h^{m-1} \cdot \left(\frac{F}{A}\right) \cdot \left(\frac{A}{A_c}\right)^{m}$$

$$D = \tfrac{1}{2} \cdot D_0 \cdot d_h^{-q} \cdot \left(\frac{A}{A_c}\right)^{n}$$

It is easy to see how fin geometry transforms into correlation constants. Characteristics of different tube types are now a function of the same flow number, Ry.

The new heat transfer and pressure drop numbers are directly correlated:

$$Ny^{n/m} \cdot C^{-n/m} \cdot D = Ey, \qquad \text{or} \quad Ny^{\kappa} \sim Ey.$$

For turbulent pipe flow $\kappa \approx 3$. This ratio varies from 3 to 5 for a wide range of geometries. As a general rule - the more turbulent the flow the lower the exponent ratio. Typically, a single row system with a channel type core tube and basically laminar air flow will be in the range of 5.

4 Base Evaluation

If the ratio of κ is approximately identical for two systems a first simple comparison may be made by using the direct correlation form. Note that any test point of the two systems may be used. It is not even necessary to know the effective air velocity. The basic procedure is to find out if the extension of the operation point along the performance curves of system "1" matches system "2". For pressure drop we find

$$Ey_2 \overset{?}{\approx} Ey_1 \cdot \left(\frac{Ny_1}{Ny_2}\right)^{\kappa}$$

The system with the lower pressure drop number is in tendency the better one. However, it must be considered that the correlation constants imply different hydraulic length factors (included in the correlation constants) which may shift the evaluation. Also the comparison considers only the core correlations and excludes external arrangement effects. No consideration is made for total fan power. Last not least the effect of different air temperature rise which defines the exchanger mean temperature difference and thus, total heat duty is not considered.

5 General Evaluation

A meaningful evaluation must be based on an effective design case. Therefore, we take the design of a large air-cooled steam condensing system (ACC) as benchmark. A logical candidate for optimum is the fin tube system requiring the lowest face (plot) area. In that case we find small bundle and secondary cost such as fan size, steelwork or connection piping. In the following we shall use the approach as outlined in [2] and used in standards [3] and [4].

The comparison is made at same design conditions for identical heat duty and pumping power. Two systems must be known using the new form of correlations:

System 1: $\qquad Ny_1 = C_1 \cdot Ry_1^{m_1}; \qquad Ey_1 = D_1 \cdot Ry_1^{n_1}$

System 2: $\qquad Ny_2 = C_2 \cdot Ry_2^{m_2}; \qquad Ey_2 = D_2 \cdot Ry_2^{n_2}$

Note that in the following procedure average air side properties are assumed to be constant so that $Ry \sim u$, $Ny \sim h_A$ and $Ey \sim \Delta P$.

In a first step the pumping power will be calculated.

$$\dot{N} = \frac{1}{\varepsilon_F} \cdot u \cdot A \cdot \Delta P_t$$

Possible variation of fan efficiency will not be considered. At this point keep in mind that in both designs air speed and total face area are not identical.

Total air side pressure drop is a combination of bundle pressure drop and parasitic loss which depends only on external arrangement and is generally taken as quadratic function of air speed. Assuming fraction ζ of system „1" bundle pressure drop as parasitic loss the relative total pressure drop ratio will be

$$\frac{\Delta P_{t,2}}{\Delta P_{t,1}} = \frac{\Delta P_2 + \zeta \cdot \Delta P_1 \cdot (u_2/u_1)^2}{\Delta P_1 + \zeta \cdot \Delta P_1}$$

With the abbreviations defined in the annex we find

$$\frac{\Delta P_2}{\Delta P_1} = p_1 \cdot r^{n_2}$$

$$\frac{\Delta P_{t,2}}{\Delta P_{t,1}} = \frac{p_1 \cdot r^{n_2} + \zeta \cdot r^2}{1+\zeta}$$

$$\frac{\dot{N}_2}{\dot{N}_1} = \frac{u_2}{u_1} \cdot \frac{A_2}{A_1} \cdot \frac{\Delta P_{t2}}{\Delta P_{t1}} = r \cdot \alpha \cdot \left(\frac{p_1 \cdot r^{n_2} + \zeta \cdot r^2}{1+\zeta} \right)$$

For same fan power this leads to an implicit equation of flow number ratio.

$$r \approx \left(\frac{1}{\alpha} \cdot \frac{1+\zeta}{p_1 + \zeta \cdot r^{2-n_2}} \right)^{1/(1+n_2)} \tag{4}$$

Face area and air speed are connected via heat duty balance:

$$\dot{Q} = \eta \cdot c_P \cdot Ry \cdot A \cdot \left(1 - e^{-NTU}\right) \cdot ITD$$

with
$$NTU = \frac{U \cdot F}{c_P \cdot \rho \cdot u \cdot A} = C \cdot \frac{U}{h_A} \cdot Pr^{-\frac{2}{3}} \cdot Ry^{m-1}$$

The fin side coefficient governs overall heat transfer with ACC's. So, simplifying we avoid complexities of internal heat transfer calculation and take $U/h_A \approx 0{,}85$ approximately constant so that

$$NTU_2 / NTU_1 \cong s_1 \cdot r^{m_2-1}$$

For same heat duty we find the relative face area ratio as

$$\alpha \cong \frac{1 - e^{-NTU_1}}{1 - e^{-NTU_1 \cdot s_1 \cdot r^{m_2-1}}} \cdot r^{-1} \tag{5}$$

or, by re-arrangement of (4) and (5)

$$r \approx \left\{ -\frac{\zeta}{p_1} + \frac{(1+\zeta)}{p_1} \cdot \frac{1 - e^{-NTU_1 \cdot s_1 \cdot r^{m_2-1}}}{1 - e^{-NTU_1}} \right\}^{1/\kappa \cdot m_2} \tag{6}$$

Simplifying, face area ratio α may be interpreted as inverse cost ratio of the two systems. This is because face area is directly related to plot area which on its part defines overall cost. And of course, cost minimum (i.e. face area minimum) will be the optimum in practical applications. In this form the comparison needs only one operation point at the same air speed apart from flow number exponents.

6 Optimization Examples

To evaluate equations (5) and (6) apart from the correlation constants we need reference design data. Apart from air speed the *NTU* number is relevant for design – see [5]. For the following general discussion data from practical experience will suffice. Typical values are

- Multi row systems: $\quad NTU_1 \approx 1.1; \quad Ry_1 \approx 1.7 \cdot 10^5 \left(\frac{1}{m}\right); \quad m \approx 0.4; \quad \kappa \approx 4$

- Single row systems: $\quad NTU_1 \approx 1.4; \quad Ry_1 \approx 1.3 \cdot 10^5 \left(\frac{1}{m}\right); \quad m \approx 0.25; \quad \kappa \approx 5$

The constants have been selected to match the pre-dominantly turbulent flow regime of multi-row fin tube systems ($m \approx 0.4\text{-}0.5$) while single row systems tend to be laminar ($m \approx 0.2\text{-}0.3$) as a consequence of the elongated air flow path between the fins. The corresponding pressure drop exponents will be estimated by exponent ratio κ.

We shall consider the system with less face area as thermo-hydraulically superior.

As a first example see fig. 3 showing the dependency of face area from relative pressure drop and heat transfer coefficient of pre-dominantly turbulent (i.e. multi-row) flow systems. Operation points above face area factor "1" indicate poorer thermohydraulic performance. All points crossing face area factor "1" will be equivalent.

This means that a system producing 60% higher pressure drop at same face area must be at least 27% better in heat transfer to be equivalent. At 160% pressure drop and only 110% heat transfer the required face area increases by 10%. Economically, this may only be made up for by reducing face area system cost by at least 10%. The main reason for face area enlargement is the reduction of air speed forced by fixed fan power consumption (fig. 4).

The situation is worse in less turbulent (laminar) flow regimes – see fig. 5. At 160% pressure drop and same face area the required heat transfer must be enlarged by at least 47% - a value not attainable by "normal" geometry measures. Addition of turbulence promoters or addition of extra fin tube surface will generally not do the job.

This result discloses the fundamental impact of pressure drop on exchanger design and fin tube optimization. To enhance thermohydraulic performance of fin tube systems it must be ensured that turbulence promoting devices or extra surfaces do not have a disastrous effect on pressure drop. This is exemplified by different elliptical core fin tube systems. If the finning is rectangular with pushed-on fins spacers to keep the proper fin pitch are needed (fig. 1c). These spacers act at the same time as turbulence promoters which add to heat transfer as well as pressure drop. However, if the fins are elliptically wound around the core tube no turbulence promoters are required and the pressure drop is considerably lower. In practice, non-turbulence-promoted wound fin tube systems require much less fin surface than rectangular pushed-on variants for same design conditions. The selection of the latter arrangement may only be justified by considerably lower manufacturing cost.

The challenge to augment fin tube performance is generally higher with less turbulent (i.e. laminar) flow tube types. This is because the lower flow number exponent leads to smaller changes of heat transfer at air speed variation. Total fan power restriction forces these systems to operate in unfavorably low air speed ranges.

This is especially true when environmental aspects such as noise restriction come into play – typical for contemporary ACC designs. Noise generation is primarily related to air speed and air side pressure drop. Therefore, noise restraints enforce the reduction of air speed and thus, enlargement of NTU. For the general discussion we assume reference NTU_1 to rise by 17%. As a consequence, variations of heat transfer have less effect on overall design. This is confirmed by the example of low noise turbulent flow regime designs (fig. 6). In that case the required heat transfer at 160% pressure drop rises by 35% instead of 27% as with standard conditions. So, the challenge for the developer of low noise heat transfer systems is much higher. At only 10% rise of heat transfer coefficient and 160% pressure drop we need 12% more face area instead of 10% as in the normal case. This behavior is aggravated in the case of less turbulent (i.e. laminar) flow regimes because the NTU range is even higher.

Note that these results already include an abatement caused by the assumption of parasitic pressure loss. Parasitic loss cannot be avoided because the air flow up-stream and down-stream of the heat exchangers consume some energy. It is self-understanding that parasitic loss suppression is a basic requirement of competitive cooler design - however, the ideal case of no parasitic loss is practically un-attainable.

None-the-less we take a no parasitic loss example as an extreme limit. As could be expected (fig. 7) the effect of pressure drop variation is larger in the case of no parasitic pressure loss. E.g. at 160% pressure drop the required heat transfer must rise by 30% instead of 27% as in the normal case. The reason for this result is that parasitic losses are not affected by relative pressure drop variation because they only depend on external cooler layout. On the other hand – with avoiding all parasitic loss the condition of identical fan power consumption is more easily to be met. With no parasitic loss the example of 160% pressure drop ratio reduces to effectively 112% and the required heat transfer enhancement would be down to plus 5%.

7 Summary

To optimize heat transfer systems not only heat transfer coefficients and surface density must be considered but also pressure drop and pertinent flow regime. Environmental requirements such as noise abatement restrict the options for optimization to a large extent. Boundary conditions like parasitic loss must be taken into account. Comparison of the fin tube characteristics alone will not suffice.

The presented comparison method allows to check step-by-step the effectiveness of proposed fin tube variants. The internal geometry of systems within the "black box" is not restricted as long as they can be related to air face area. It is possible to directly compare completely different geometries (such as tube diameter, fin diameter, tube pitches, fin pitches, number of tube rows, etc.).

No specific fin tube performance augmentation measure is proposed. However, the approach to simply add more fins to the inch (i.e. increasing surface density) for performance enhancement may be put into question. The same holds for adding simple turbulence promoting devices. The proportionality of heat transfer and pressure drop is dominated by the exponent ratio of the two curves. Turbulence promoters leading to over-proportional pressure drop rise are shown to be no efficient measure for thermohydraulic optimization. Last no least - the evaluation of each fin tube variant must be accompanied by cost assessment to avoid a dead end. No one will buy a "smart" cooler which is too expensive.

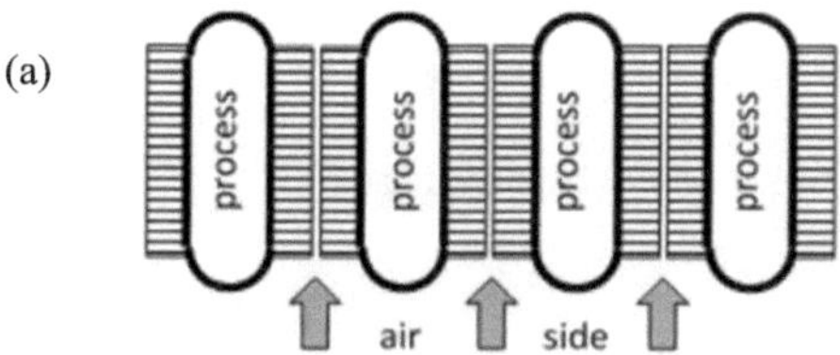

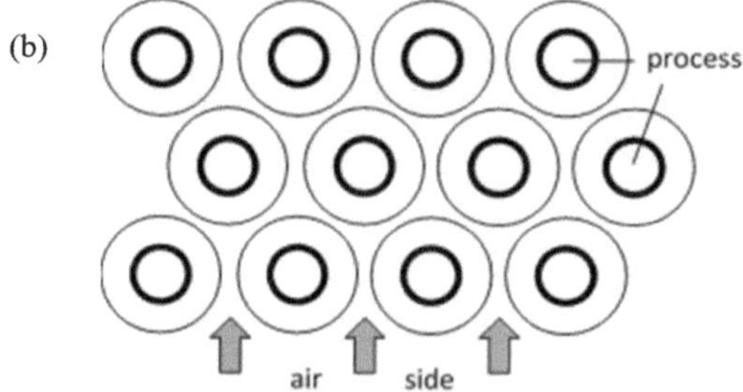

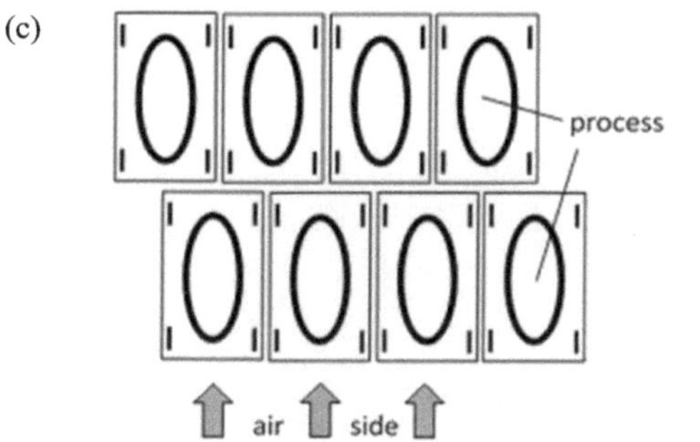

Fig. 1: Cross-flow Fin Tube Systems,
(a) single row, (b) circular multi-row,
(c) elliptical multi-row with rectangular finning

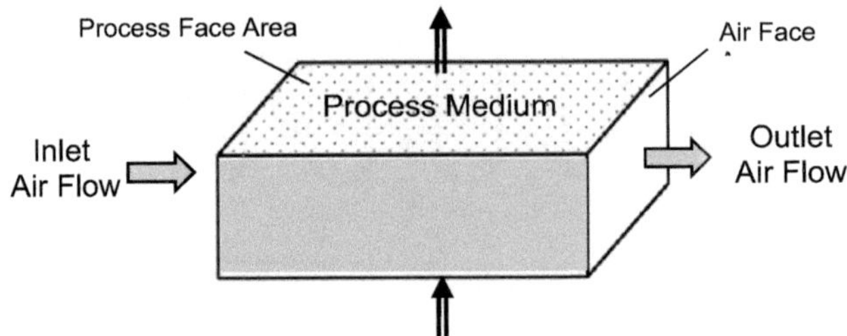

Fig. 2: Box type Heat Exchanger Model

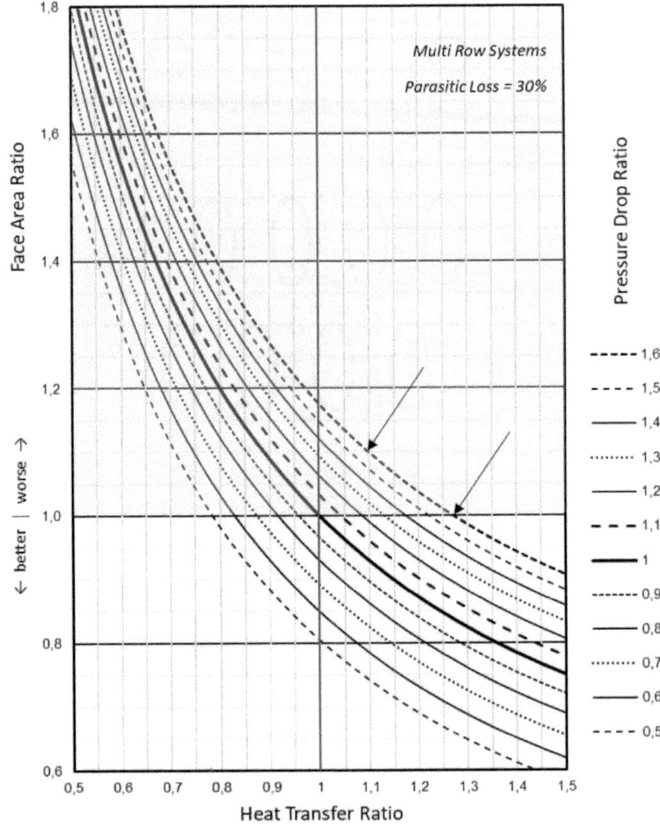

Fig. 3: Face Area of Turbulent Flow Regimes

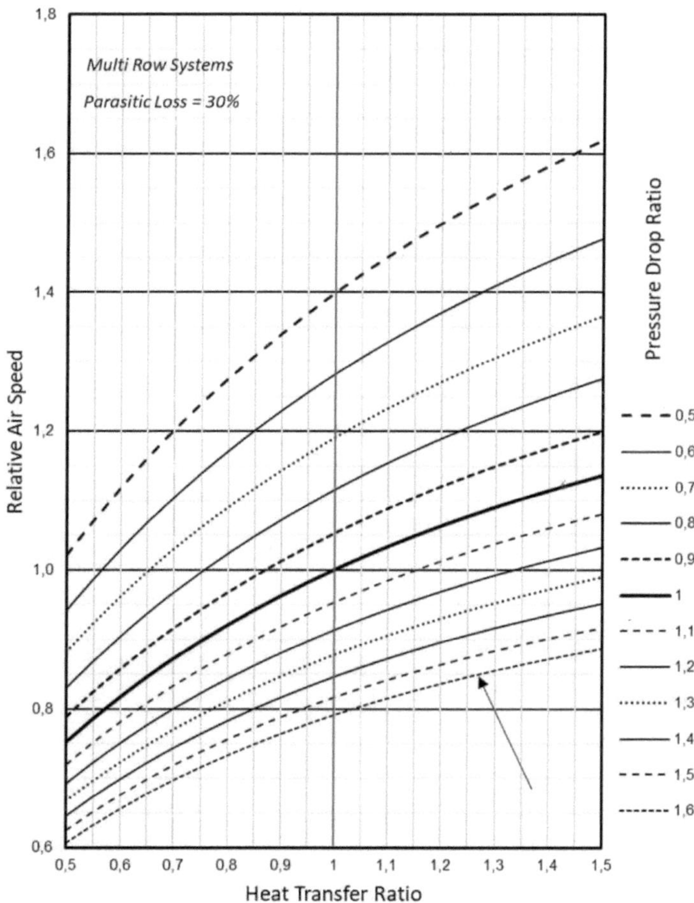

Fig. 4: Relative Air Speed of Turbulent Flow Regimes

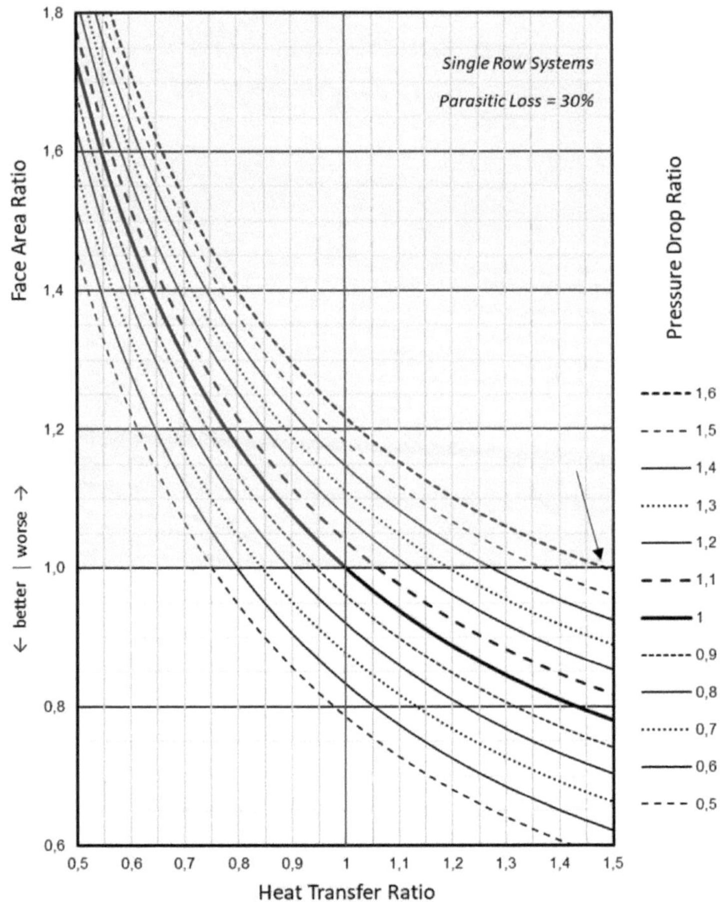

Fig. 5: Face Area of Laminarized Flow Regimes

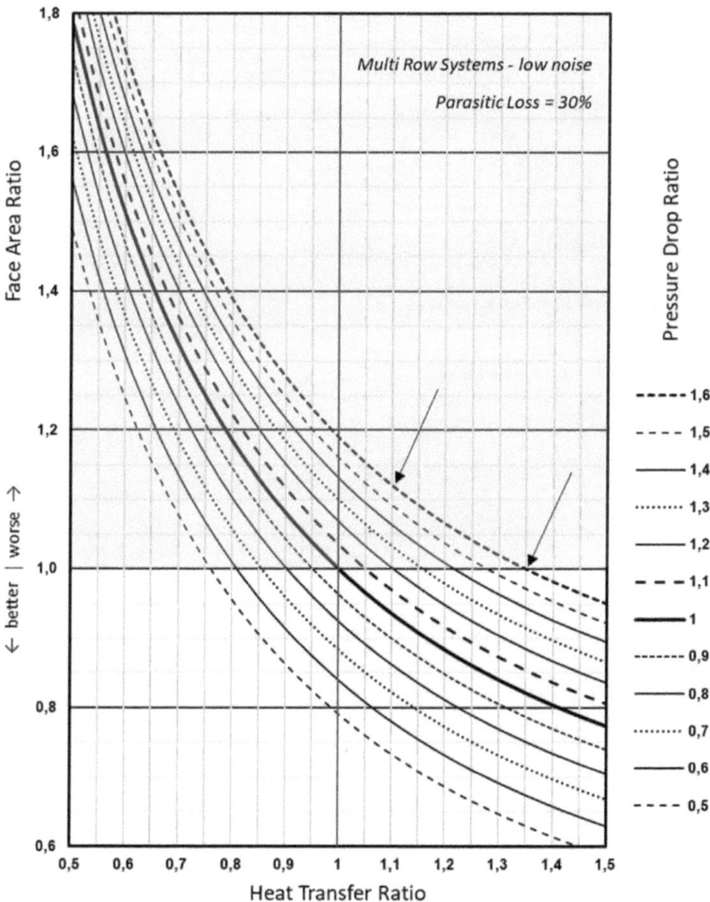

Fig. 6: Face Area of Low Noise Turbulent Flow Regimes

15

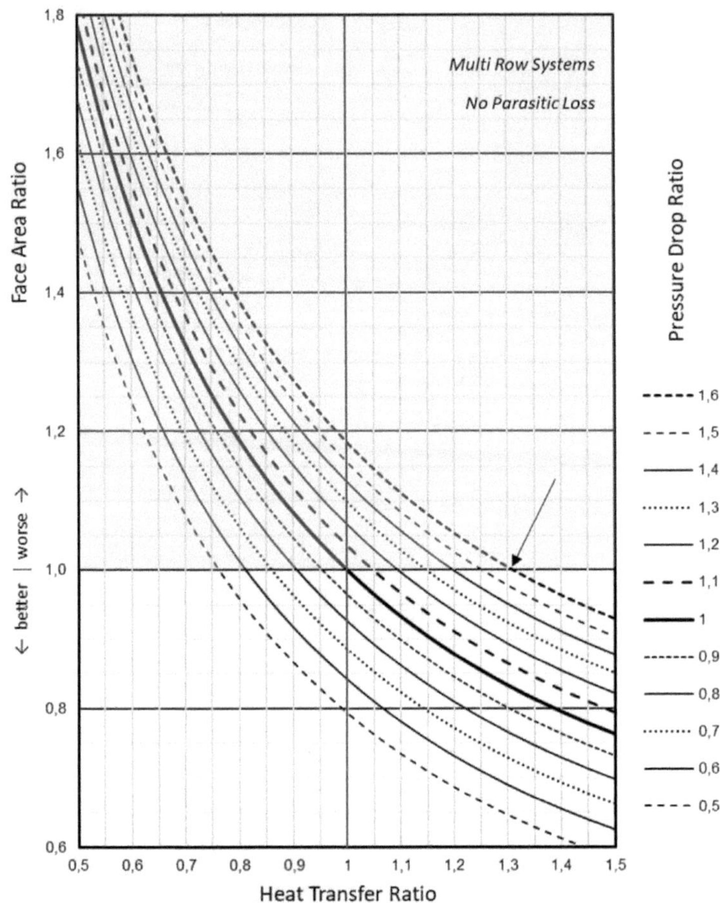

Fig. 7: Face Area of Turbulent Flow Regimes, No Parasitic Loss

Nomenclature

A	-	exchanger air face area (m²)
A_c	-	air flow area in narrowest gap (m²)
A/A_c	-	area concentration ratio
C	-	heat transfer correlation constant
c_P	-	average air side specific heat capacity (kJ/kg K)
D	-	pressure drop correlation constant
d_h	-	hydraulic diameter (m)
Ey	-	pressure drop number (1/m²), pseudo Euler number
F	-	total air side (finned) surface (m²)
F/A	-	surface density
l_h	-	hydraulic air flow length (m)
h_A	-	fin tube heat transfer coefficient (W/m²K)
ITD	-	initial temperature difference process-air side (K)
m	-	heat transfer flow number exponent
$n = 2 - q$	-	fin tube pressure drop flow number exponent
$Nu = h_A \cdot d_h/\lambda$	-	Nusselt number (-)
NTU	-	number of transfer units (-)
Ny	-	heat transfer number (1/m), pseudo Nusselt number
Pr	-	air side Prandtl number (-)
$p_1 = Ey_2(Ry_1)/Ey_1$	-	fin tube pressure drop ratio for same speed Ry_1
q	-	friction factor Reynolds exponent
$\dot{Q}$	-	heat duty (W)
$Re = \rho \cdot u_c \cdot d_h/\eta$	-	air side Reynolds number
$r = Ry_2/Ry_1$	-	flow number ratio
Ry	-	flow number (1/m), pseudo Reynolds number
$s_1 = Ny_2(Ry_1)/Ny_1$	-	fin tube heat transfer ratio for same speed Ry_1
u	-	air face velocity (m/s)
u_c	-	air speed in narrowest gap (m/s)
U	-	overall heat transfer coefficient (W/m² K)
$\alpha = A_2/A_1$	-	face area ratio

ΔP	-	fin tube system pressure drop (Pa)
ΔP_t	-	total air side pressure drop (Pa)
ρ	-	average air side density (kg/m³)
η	-	average air side viscosity (Pa s)
λ	-	average air side conductivity (W/m K)
$\varsigma = D_0 \cdot \mathrm{Re}^{-q}$	-	fin tube resistance coefficient
ζ	-	parasitic pressure loss fraction
ε_F	-	fan efficiency
$\kappa = n/m$	-	exponent ratio

Subscripts:

0	-	referring to classical correlation
1	-	referring to system 1
2	-	referring to system 2
c	-	referring to narrowest gap

Bibliography

[1]   Detlev G. Kröger (Department of Mechanical Engineering, University of Stellenbosch) (1998) Air-Cooled Heat Exchangers and Cooling Towers. Begell House

[2]   Hans G. Schrey (1996) Zum Leistungsnachweis luftgekühlter Kondensatoren. Brennstoff-Wärme-Kraft 9, paper 825, VDI Düsseldorf (in German)

[3]   VGB R-131 Me (1997) VGB Guideline Acceptance Test Measurements and Operation Monitoring of Air-Cooled Condensers under Vacuum. VGB Technische Vereinigung der Grosskraftwerksbetreiber

[4]   ASME PTC 30.1 -2007 (2008) Air-Cooled Steam Condensers, Performance Test Codes. The American Society of Mechanical Engineers

[5]   Hans G. Schrey (2001) Practical Aspects of Air-Cooled Exchanger Design. 12th IAHR Cooling Tower & Spraying Pond Symposium, UTS Sydney, Australia

YOUR KNOWLEDGE HAS VALUE

- We will publish your bachelor's and master's thesis, essays and papers

- Your own eBook and book - sold worldwide in all relevant shops

- Earn money with each sale

Upload your text at www.GRIN.com and publish for free